# Weather Whys

## Second Edition

## Questions, Facts, and Riddles About Weather

### By Mike Artell

Good Year Books

# Good Year Books

are available for most basic curriculum subjects plus many enrichment areas.
For more Good Year Books, contact your local bookseller or educational dealer.
For a complete catalog with information about other Good Year Books, please contact:

Good Year Books
P. O. Box 91858
Tucson, AZ 85752
www.goodyearbooks.com

Book design by Christine Ronan.
Additional pages designed by Dan Miedaner.

Photo Credits:
Page 14, buoy: NOAA
Page 36, satellite image of California: NOAA
Page 37, California fires: U.S. House of Representatives, Committee on Resources
Page 88, Hurricane Floyd – flood image: NASA/GSFC
Page 89, Buffalo Bayou: Harris County Flood Control District
Page 90, normal and chapped lips: U.S. National Library of Medicine and
the National Institutes of Health

This book is filled with information about weather, information that is based on the things children want to know and which anticipates the "why" questions children ask every day. But this book is not just a collection of facts and figures about weather. Each page also contains hilarious cartoons, ridiculous riddles, and easy-to-do activities which help bring the material to life in a way that children find interesting, exciting, and fun.

If you know a child who is curious about clouds, snow, tornadoes, thunderstorms or any weather phenomenon, this book can be a wonderful companion. And you'll enjoy it, too!

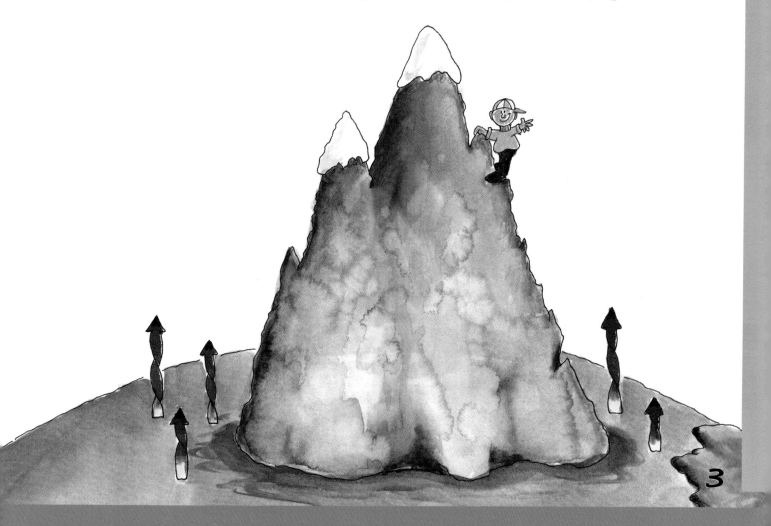

# Table of Contents

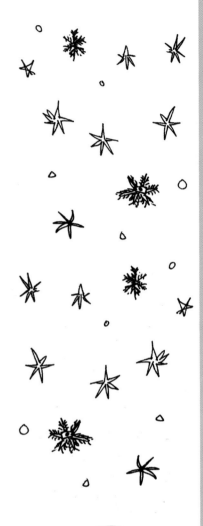

# Why do we need to know about the weather?

Look around. Weather is everywhere! Weather is what we call a lot of different things that make up what it's like outside.

When we say the weather is "bad," we might mean it's freezing cold with sleet falling. Or we might mean that it is raining hard and the wind is really blowing. When we say the weather is "good," we might mean that the sky is clear and the sun is shining. Or we might mean that it's nice and cool with just a little breeze blowing.

Actually, the weather itself isn't good or bad. It just depends what kind of weather you like. We enjoy being out in "good" weather. We stay inside when the weather is "bad."

Weather has a lot to do with the way we dress, the food we grow, where we live, and how we travel. All these things depend in some way on what is going on in the air around us, and that's what weather is all about. Because weather plays such an important part in our lives, it's a good idea to understand a little about how the weather works. A good place to learn about weather is during the weather report on TV news each night. Often, the man or woman who talks about the weather is a meteorologist.

## METEOROLOGIST

WHOA! BiG WORD!

A meteorologist is a man or woman who is trained to understand and predict the weather. Because weather changes so fast, meteorologists sometimes make mistakes, but most of the time they are very accurate.

7

# More about weather

When you watch the weather report on TV, you'll learn how hot or cold it was today, how much rain fell, and what the weather is probably going to be like tomorrow.

You'll also see pictures taken by satellites high above the earth showing clouds moving across the land and water.

MAVIS

STUFF YOU CAN DO!

You can **learn about weather** by reading books in the library and by trying some of your own weather experiments.

You can **place a thermometer in a shady spot** and write down the temperature every day at certain times.

You might even want to **make your own weather vane** to show the direction the wind is blowing.

You can **catch rain water** in a cup and measure how much rain fell in a day or a week.

There's another place you can go to learn about the weather. You can ask an adult you know. Farmers, construction workers, fishermen, pilots, and truck drivers all watch the weather very carefully. In fact, anyone who works outside or travels a lot is very interested in what the weather is going to be. You can ask them why the weather is important to them. They'll probably have lots of interesting stories to tell you about the weather.

# **Why** do we care what the weather is going to be?

Knowing what the weather is going to be makes it easier to plan outdoor activities. For example, if there is a good chance that it is going to rain, you won't want to plan a picnic.

I once knew a cow I called Heather

who didn't know much about weather.

Once out in the rain,

she felt a strange pain,

when she found out that water

shrinks leather.

Knowing what the weather will be helps us avoid danger, too. This is especially true if a hurricane is on the way. A hurricane is a very strong wind storm that can blow down buildings and hurt people. Usually, there are very high waves and strong winds near the seashore when a hurricane is in the area. People who live near the shore take their important things and their pets and move away from their homes until the hurricane passes by. Knowing that a hurricane is on its way can save lives.

# More about weather forecasting

## METEOROLOGIST'S TOOLS

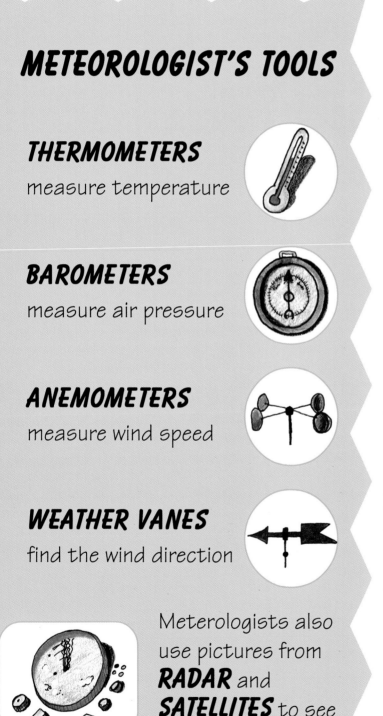

### THERMOMETERS

measure temperature

### BAROMETERS

measure air pressure

### ANEMOMETERS

measure wind speed

### WEATHER VANES

find the wind direction

Meterologists also use pictures from **RADAR** and **SATELLITES** to see how the weather is changing.

People called meteorologists have many ways of figuring out what the weather will be like. This is called forecasting.

You don't have to be a meteorologist to predict all weather changes. Sometimes you can see a change coming if you look at the sky. Clouds that are loose and puffy usually mean that the weather will be good. But when the puffy clouds start piling up higher and higher and getting dark, it usually means stormy weather is on the way.

"These plants really need water."

"Hey! Here comes some rain!"

The color of the sky can also help weather forecasters. If the sky looks red in the evening when the sun goes down, it usually means the weather will be nice the next day.

If you like to grow plants outside, you probably care about the weather. Plants need the right temperature and the right amount of sunlight and water to grow. If the weather is dry for a while, you might have to water your garden. If your plants aren't getting enough sunlight, you may have to move your garden to a sunnier spot. If the weather gets too cold or hot, you may have to cover your plants to protect them.

# What is a Buoy?

A buoy is a marker that floats in one spot on the surface of a body of water. It is usually brightly colored and easy to see. Buoys are used to show sailors where it is safe to sail a boat; those buoys usually have lights or bells.

Weather scientists use buoys too. They put sensors on the buoys that gather lots of information about the weather. This information includes:

- Air pressure
- Wind direction and speed
- Temperature of the air and sea
- The size, shape, and direction of waves

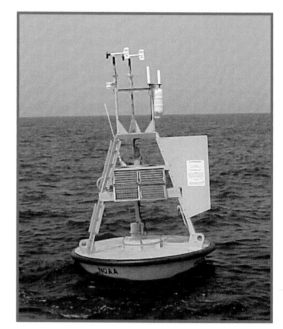

Buoys like this one gather information about weather conditions at sea.

The buoys radio the information to a station, where scientists can study it. Buoys can be dropped into the ocean from airplanes or they can be lowered into the water from a boat.

# The WRONG Name

Even though weather forecasting gets more and more accurate, it is still very hard to predict exactly what the weather is going to do. It is not uncommon for weather forecasts to be wrong.

When the civilian national weather service was created in 1890, someone suggested that the name be the "**W**eather **R**eporting **O**ffice of the **N**ational **G**overnment." Thank goodness they decided not to use that name. If they had, the agency would have been known as **WRONG.**

# NOAA

NOAA isn't the name of a person. It's the name of an organization that warns us about dangerous weather. The letters NOAA stand for National Oceanic and Atmospheric Administration. NOAA was created on October 3, 1970, to gather information about the environment and to use that information to help protect people and property from natural disasters.

**Here are some of the things NOAA workers do:**

- They fly into the eyes of hurricanes to gather information about where a hurricane might hit land.
- They help free whales that are trapped in ice.
- They help clean up oil spills.
- They watch the movement of ash from volcanoes and wildfires and warn us if the air is unsafe.

# Why can't we see air?

We can't see air because the gases that make up air are too small. Actually, the molecules that make up those gases are too small.

"Wait a minute! Let's start from . . ."

# THE BEGINNING

"Nyah, nyah. You can't see us."

Air is a combination of many gases, but mostly nitrogen gas. In fact, if you could put 100 little bits of air in a jar, about 78 of them would be nitrogen gas. Oxygen would make up most of the rest. These little bits of gas, called molecules, are so small, it's impossible to see them with your eyes alone.

16

All living things need air to live. That's because we all have to breathe in some way. Human beings and many animals have lungs which help move air into their blood. Lungs are like air bags inside your chest.

"You don't look so good."

Fish don't have lungs—they have gills. Gills are special organs that capture the air in the water. Yes, that's right: there's oxygen in the water, too. Our lungs aren't designed to capture the air in water so we can't breathe underwater. We have to hold our breath.

# More about air

We also need air to make fires. In fact, one of the ways people put out fires is to throw dirt on them. This blocks the air and causes the fire to go out.

Sometimes the air smells funny, but air really doesn't have a smell of its own. What we smell are particles of other things in nature that are carried through the air.

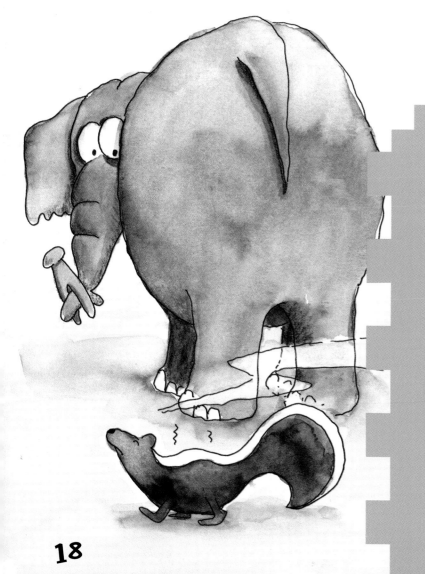

An elephant said to a skunk,

"Would you please stand away from my trunk?

When you entered the room, I smelled your perfume,

and to tell you the truth, skunk, it stunk!"

When we breathe in, we take air into our lungs. When we breathe out, we blow out air along with another kind of gas called carbon dioxide. Too much carbon dioxide is not good for people. We would be in trouble if there was too much carbon dioxide in the atmosphere. Fortunately, plants breathe in carbon dioxide and breathe out oxygen. This keeps the balance of air and carbon dioxide just right.

**"We're breathing buddies!"**

# AIR

Here are just a few ways people use air (you can probably think of more):

To blow up balloons

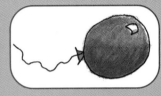

To lift airplanes

To make fires burn

To whistle

19

# Why is it hard to breathe at the top of high mountains?

"Puff, puff"

There is much more air near the surface of the earth than there is at the tops of high mountains. Even though air molecules are very light, they still have some weight. As the air molecules pile up on each other, most of them get pushed down toward the surface of the earth. As they fall to earth, the weight of the air gets heavier and heavier.

Hey . . . quit pushing!

20

# BACK IN THE 1600s

Hundreds of years ago, a scientist named Evangelista Torricelli invented an instrument to measure the weight of the air. This instrument is called a barometer.

Barometers are very helpful in forecasting the weather. When a barometer shows that the air pressure is low, it usually means that bad weather is coming. When the barometer shows high air pressure, it usually means the weather will be good.

# More about air pressure

Many years ago, people dreamed of flying through the air. They knew from watching the weather that warm air rises, so they built giant hot-air balloons with big baskets in which people could ride.

## Little Riddles

**Q.** What part of a mountain can see the best?

**A.** The peek.

The air inside these balloons was hot enough to make them rise thousands of feet into the sky. Hot-air balloons are not the only kind of balloons that can fly.

Have you ever gotten a balloon at a carnival or fair that floated in the air by itself? Well, there's a special gas called helium inside that balloon that is even lighter than air. That's why it floats. What do you think would happen to that balloon if you let go of it? At first, it would fly very high into the air. But as the balloon got farther and farther from the earth, there would be less and less air around it. Soon the helium inside would push so hard it would pop the balloon.

# Why do we have to protect ourselves from the sun's rays?

Even though the sun looks small in the sky, it's actually many, many times bigger than the earth. The reason it looks small is because it's so far away—about 93 million miles from earth. If you could get close to the sun, you'd see that it is a gigantic fire ball sending huge flames thousands of miles into space. It is so bright that it lights up the whole side of the earth facing it. This is great because it keeps our planet warm, but those rays can also burn your skin and hurt your eyes.

Uh oh!

There are other rays from the sun that we can't see. Some of these, called ultraviolet rays, can be very harmful to people. If you've ever been sunburned, you know about ultraviolet rays. Those are the rays that burn your skin and make it sore to touch.

The sun's ultraviolet rays can also be very harmful to your eyes, so be sure to wear sunglasses. Looking directly at the sun can burn the retina in your eye. Once this retina is burned, it is damaged forever. So don't look directly at the sun. Ever.

If you go out in the sun, it's a good idea to cover the parts of your body that might get sunburned. Sometimes you can cover your skin with a long shirt or jeans, or you can use a special lotion that protects your skin from the sun.

# IMPORTANT INFORMATION

**Never, never, NEVER look directly at the sun!**

# More about ultraviolet light

There's one more thing to remember. Just as the sun warms the earth, it also warms people and animals. When you're outside and you feel yourself getting hot, it's a good idea to stop and cool off. You can sit in the shade or just stop and rest. It's also a good idea to drink lots of water when the weather is warm. Our bodies need more water when we get hot, and the water helps us keep cool. If you have a pet that stays outside, make sure you keep its water dish filled during warm weather too.

# SOLAR ECLIPSE

The sun, the moon, and the earth are always moving through space. Sometimes, the moon will move between the sun and the earth. When this happens, part of the sun is blocked out and we have a solar eclipse.

This doesn't happen very often, so people usually want to go outside and watch it. If you want to watch a solar eclipse, remember: DO NOT LOOK DIRECTLY AT THE SUN.

## STUFF YOU CAN DO!

If you want to watch an eclipse, make a tiny hole in a piece of construction paper. Turn your back to the sun and hold the construction paper as if you were reading words on it. With your other hand, place a piece of white paper about 12 inches below the construction paper. You will see the eclipse perfectly, and you won't hurt your eyes.

27

# **Why** is the temperature colder at the top of a mountain than at the bottom?

As the sun shines down on the earth, it warms the land and water. If you've ever walked barefoot on sand or a sidewalk in the summer, you know that the sun can make things hot. Some of this heat bounces back into the air, but the farther you get from the earth's surface, the less heat you feel.

EEE!

AAA!

OOO...

## FAST FACT

You don't have to be at the top of a mountain to find cold weather. In 1983, at the Russian station Vostok in Antarctica, scientists recorded the lowest temperature ever: −128.6°F.

If you start climbing a mountain, you'll be moving away from the surface of the earth. You won't notice much temperature difference at first, but soon it will get cooler. That's because less and less heat from the earth is reaching you. If the mountain you're climbing is very tall, it may be very cold at the top. Some mountains are so tall, they are covered in snow and ice all year long.

# More about temperature

There's another reason why it may be colder at the top of a mountain. As warm, wet air rises from the earth's surface, it creates clouds. The wind may blow the clouds up against the mountains, making the clouds "pile up." When this happens, the clouds become heavy with water vapor and drop the water onto the mountains as rain or snow.

As the clouds lose their water, the air around them becomes colder. Anyone near the top of the mountain would feel colder than anyone near the base of the mountain.

When the wind blows, it can also make us feel cooler than when there is no wind blowing. In fact, if the temperature is 50°F and the wind is blowing at 15 miles per hour, it feels the same to our skin as if it was 36°F with no wind. This cooling effect of wind on our skin is called wind chill.

## Wind Chill Chart

| If the actual temperature is: | and the wind is blowing: | it feels as if it's: |
| --- | --- | --- |
| 50°F | calm | 50°F |
| 50°F | 15 mph | 36°F |
| 41°F | 25 mph | 17°F |
| 32°F | 10 mph | 18°F |
| 14°F | 15 mph | –13°F |
| –4°F | 25 mph | –50°F |

**31**

# Why is it windy on some days and calm on other days?

I knew that!

Believe it or not, wind has a lot to do with the sun!

Even though it may be calm where you are, the air is always moving around the earth. Wind begins when the sun warms the air, the land, and the water. During the day, the land warms up faster than the water, and when the air gets warm, it starts to rise. When this warm air rises, cool air moves in to take its place. When the warm air gets high enough in the sky, it starts to cool off, and when it gets cool enough, it starts to fall. Meanwhile, the cool air that moved in over the land starts to warm and it begins to rise.

This warming and cooling of the air causes lots of air movement, and moving air is what we call wind. At night, just the opposite happens. When the sun goes down, the land cools off more quickly than the water. Then, the air over the water starts to rise and that makes the cooler air over the land move in to take its place.

**"All this wind is making me dizzy!"**

# More about wind

Next time you go to the beach on a warm day, notice which way the wind is blowing. During the day, it's probably blowing from the water onto the land. At night, it's probably blowing from the land onto the water. Even though we can't see the wind, we can see what it does. The wind can fly kites, turn windmills, push sailboats, make flags flap, pollinate plants, and do a hundred other things.

**FAST FACT** Some places on earth are very windy. The windiest place on earth is the George V Coast on the continent of Antarctica. The winds there have been measured at more than 200 miles per hour.

Not all wind is near the surface of the earth. Some very powerful winds called the jet stream are more than 6 miles above the earth. Depending on which way an airplane is going, jet stream winds can slow down or speed up airplanes that fly through them.

# The Santa Ana winds

Santa Ana winds form when the air pressure builds up in the high desert areas west of the Rocky Mountains. The air gets pushed down the western slope of the plateau. This pushing causes the air particles to squeeze together and heat up. As the air gets hotter, it also gets drier.

During the cooler months of the year, Santa Ana winds strike the area near Los Angeles, California. These very strong winds are named after the Santa Ana Canyon through which they blow. In order for a wind to even be considered a "Santa Ana," it has to move at more than 25 knots. An average speed for a Santa Ana is 35 knots and sometimes the winds can gust as high as 100 knots.

Satellite image of California wildfires showing Santa Ana winds blowing smoke.

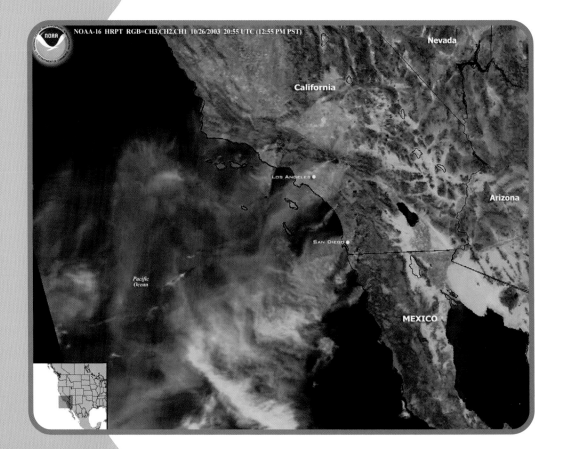

36

The Santa Ana winds swoop through the canyons and over the plants and trees. As they blow over the leaves and branches, much of the water in the plants evaporates. This creates large areas of dry, brittle wood, and that raises the danger of wildfires.

"Santa Ana winds can make the air hot and dry, creating perfect conditions for forest fires."

If a wildfire does get started, it has lots of fuel to keep it going. When the hot, dry Santa Ana winds fan the flames, it makes the fire very difficult to contain. If the wildfires get big enough, they can even affect local air currents and make it difficult for pilots to fly their aircraft.

# Why do some places have hurricanes and others don't?

Large bodies of warm water are the most likely places for hurricanes to form, so the land areas close to large bodies of warm water are the ones that usually get hit by hurricanes. In fact, hurricanes won't even form during certain times of the year or in certain places because the water temperature is too cool.

Over warm water, large thunderstorms can grow and get very powerful. Sometimes they get so large, the turning of the earth makes them start to spin. If the storm stays over warm water long enough, the winds inside of it can start spinning very fast. When they spin at 74 mph, the storm becomes a hurricane.

In order to keep track of the different hurricanes that develop each year, meteorologists started giving them names. Now, the first hurricane of the year is given a name starting with the letter "A." The next hurricane is given a name starting with "B," and so on.

"They'll never get to me."

Z

# More about hurricanes

The center of a hurricane is called the eye. Winds near the eye of some hurricanes have reached speeds of up to 200 mph.

The air pressure inside a hurricane is usually very low. This, along with the forward movement of the hurricane, pushes water and creates a storm surge. In a storm surge, water builds up like a wall in front of the hurricane. As the hurricane reaches land, this wall of water pushes onto the shore. Anything in its path is washed away, even houses, buildings, and cars.

Today, weather forecasters watch the movement of hurricanes very carefully. By using airplanes and satellites, they know exactly where a hurricane is and where it will probably go. Sometimes they know days ahead where a hurricane is headed, and they can warn people to leave the area.

 The worst hurricane ever recorded was Hurricane Andrew. Andrew hit Southern Florida in September 1992 and destroyed billions of dollars of houses, cars, and buildings. Thousands of people lost their homes and property, and some lost their lives.

# Why are tornadoes so dangerous?

Tornadoes are dangerous for two reasons: first, they are very powerful, and second, they are very unpredictable. This means that it's very difficult to try to guess when a tornado is going to form and where it is going to go.

Maybe I'll go this way . . .

No . . . that way!

Hundreds and hundreds of tornadoes are reported each year from every state except Alaska. Tornadoes do occur in Alaska, but that's very rare.

ALASKA

When tornadoes form, there's almost always a heavy thunderstorm nearby. Most of us know a tornado from the shape of its cloud. These clouds are called funnel clouds because they're shaped much like a funnel that you might use in the kitchen.

# More about tornadoes

In order for a tornado to do any real damage, it has to touch something on the ground. Many times the tip of a tornado will touch a spot on the ground for few seconds and then lift up. It will then move and touch the ground again some distance away. Whatever the tip of the tornado touches while it is on the ground is sure to be badly damaged. Both tornadoes and hurricanes can destroy homes and property. But, unlike hurricanes, most of the damage tornadoes do is in a very narrow area. A tornado may destroy one home and not touch the home next to it. Tornadoes are very unpredictable.

Tornadoes can form over water as well as land. Tornadoes that form over water are called water spouts. Sometimes when it's raining very hard, people will say, "It's raining cats and dogs." Although dogs and cats may not really fall from the sky, water spouts have actually picked up fish from the water and dropped them on to the land miles away.

Heads up!

## Little Riddles

**Q.** What do tornadoes become when they go to sleep?

**A.** Torna-doze

z-z-z-z

# Why are there different kinds of clouds?

Clouds form when tiny droplets of water vapor rise into the air. We can't see these little droplets by themselves, but when they start collecting together, we see them as clouds. The amount of water in the air, the temperature of the air, and the wind all play a part in making and shaping clouds. There are three basic kinds of clouds: cirrus, stratus, and cumulus.

Cirrus clouds are the wispy, white, feathery clouds that are very high in the sky. Sometimes people call cirrus clouds "mare's tails" because they look like the tail of a horse. When cirrus clouds start to get thicker, it can often mean that the weather is going to get warmer.

**FAST FACT**

Cirrus clouds are high clouds. Stratus clouds are low clouds.

Stratus clouds look like layers of clouds. They are usually low in the sky and can sometimes give us rain or snow. The word stratus comes from another word which means "to lay down or spread out."

What clouds never make jokes?

The Cirrus (serious) ones

Hey! That's not funny!

47

# More about clouds

The cumulus clouds in the sky

can look like a sheep or a pie.

They change very fast,

and their shapes never last,

so you only have time to wave, "bye."

The third basic kind of cloud is called cumulus. Cumulus clouds are puffy clouds. Have you ever watched clouds in the sky that looked like animals or faces? If so, you were probably looking at cumulus clouds.

Hey! That looks like Aunt Gerty!

Next time you see cumulus clouds in the sky, write down how many animals or faces you see in the clouds. What do you think makes the clouds change shape? Could it be the wind? Cumulus clouds usually don't bring rain, but they can turn into cumulonimbus clouds.

Cumulonimbus clouds are big, puffy, and very tall. They are often called "thunderstorm" clouds and they can contain thousands of tons of water.

## STUFF YOU CAN DO!

**You can make your own cloud. Here's how:**

On the next cold day, blow your warm breath on a cold window or mirror and presto! The "foggy" place on the mirror is actually condensed water vapor. In other words . . . a cloud!

**49**

# Why can't we see through fog?

Fog is just a cloud on the ground, and we can't see through fog for the same reason we can't see through a cloud . . . all those billions of little water droplets get in the way.

Whenever warm air meets cold air, tiny droplets of water form. This process is called condensation. You've seen the same thing happen when you fill a glass with a cold liquid. Before long, tiny drops of water form on the outside of the glass. Some people say the glass is "sweating." What's really happening is that the glass becomes colder than the air in the room and condensation forms on the outside of the glass.

During the day, fog usually lifts and becomes a low stratus cloud.

See you later!

Condensation forms on the warm side of objects. That's why, when you breathe warm air on a window, the window becomes foggy. Your breath is warmer than the glass. The same thing happens when the air and the land are different temperatures. Tiny droplets of water form in the air near the ground. These droplets are so small and light that they float in the air. As the sun warms the droplets, some lift into the sky and form clouds.

Other droplets that have not been warmed yet stay near the ground and form fog. Fog can make it more difficult to drive cars and fly airplanes because the people driving or flying can't see as well. How well people can see the ground around them is called visibility.

# More about fog

It doesn't taste like pea soup!

Sometimes when the fog is very thick, people call it "pea soup." When there is fog in one area but a nearby area is sunny, people say the fog is "patchy."

## Little Riddles

**Q.** What do you call a cloud near the ground?

**A.** I used to know, but I fog-got!

**Q.** What fairy tale do water droplets like best?

**A.** The Fog Prince

Some places, such as the central part of California near the Pacific Ocean, have lots of fog. The Golden Gate Bridge in San Francisco is painted orange to make it easier to see when it's foggy in the San Francisco Bay area.

I know there's a bridge around here somewhere!

Tongue Twisters

Frogs find fine flies in the fog.

# Why do I see lightning before I hear thunder?

During a thunderstorm, you see lightning before you hear thunder because light travels faster than sound. The bright flash the lightning makes travels 186,000 miles in one second. Thunder, the sound the lightning makes, travels 1,100 feet in one second.

**"Why do I hear thunder at all?"**

*Good question!*

Lightning bolts can reach temperatures five times hotter than the surface of the sun. When lightning strikes, the lightning bolt causes the air around it to heat up very quickly. This causes the gases in the air to expand so fast that they make an exploding sound. That sound is what we call thunder.

YEEOW!

**FAST FACT** Lightning strikes somewhere on the earth about 100 times every second.

54

# What makes lightning?

Lightning is caused by tiny particles in the clouds rubbing together creating electricity. This electricity flashes down to the ground as a lightning bolt.

People in ancient Greece believed in a god named Zeus who lived on a very high mountain called Mount Olympus. They believed he would throw lightning bolts like spears down on the earth.

## Little Riddles

**Q.** How did Benjamin Franklin feel when he found out that lightning is really electricity?

**A.** He was shocked!

**Q.** What kind of insects love the sound of thunder?

**A.** Lightning Bugs

**Q.** What city gets the most thunder and lightning?

**A.** Electri-City

55

# Where does lightning strike?

Lightning usually strikes the tallest object in an area. That's why it's not a good idea to stand under a tree when lightning is around.

Maybe this isn't such a good shelter.

You might not want to stand too close.

There's an old saying that "Lightning never strikes the same place twice," but that's not true. There have been many recorded cases of lightning striking the same place many times. Lightning has even struck the same people many times. Several people have been hit by lightning more than once and have lived to tell about it.

Lots of people are frightened by the loud sound thunder makes. But thunder can't hurt you. It's just noisy.

**THUNDER MAY BE FRIGHTENING BUT THE DANGEROUS THING IS LIGHTNING**

STUFF YOU CAN DO!

If you see lightning strike in the sky, you can figure how far away it is from you. As soon as you see the flash, start counting seconds. When you hear the thunder, stop counting. Then divide the number of seconds by five. That's how many miles away the lightning struck. For example, if you saw some lightning and counted to five before you heard the thunder, you could figure that the lightning struck one mile away.

One . . .
Two . . .

# Why don't planes take off or land during bad thunderstorms?

Thunderstorms form when large amounts of warm, wet air rise and mix with cold, dry air. This forms cumulonimbus clouds, and soon a storm is born.

In thunderstorms, airplane pilots are most concerned about wind shear. Wind shear is a sudden change in the direction of the wind, especially straight up or down. In bad thunderstorms the wind can whip around in all directions. This is dangerous because airplanes depend on air moving across their wings to lift them up. If the air suddenly pushes down or sideways on the wings, the airplane will have a hard time staying up in the air.

Here's an experiment you can try. Fold a paper airplane from a piece of paper. Then go into a room that has a ceiling fan. With the fan off, toss the paper airplane across the room. Your plane probably had a nice, smooth flight.

Now, turn the fan on to the fastest speed and toss the paper airplane again, being sure that it flies a few feet beneath the turning fan. Chances are that when your airplane flew beneath the fan, the breeze from the fan quickly pushed the airplane down toward the ground. That's exactly what happens during a wind shear. Now try to imagine a room with lots of fans pointing in different directions. That's what it would be like to fly in a thunderstorm.

# More about thunderstorms

A flight simulator is a machine that helps airplane pilots practice flying in dangerous conditions such as wind shear. This teaches them what to do in case they get in trouble. But you don't need to worry about flying in an airplane in wet weather. A little rain doesn't bother airplanes. If the pilot thinks the weather is too bad, he or she will wait until the thunderstorm passes before they take off or land the airplane.

Excuse me . . . I'm trying to land

People who pilot boats watch the weather carefully, too. The National Weather Service sometimes warns pilots of small boats about winds or waves that might be dangerous to them.

Boat pilots measure wind speed in knots. A knot is a little more than one mile per hour. Usually, if the wind speed is more than 18 knots or the waves are very high, boat pilots can expect a warning about dangerous weather.

# **Why** does hail sometimes fall during a summer storm?

I feel like a yo-yo.

Hail forms in thunderstorms, and thunderstorms usually happen during warm weather. In fact, the most common month for hailstorms is August, one of the warmest months.

When ice crystals or tiny bits of dust fall through a cloud, droplets of water freeze to them. This is how hail starts. On hot days, warm air from below the cloud can rise very quickly, pushing the particles of ice and dust back up toward the colder air in the cloud. When this happens, more water freezes to the outside of the particles. This up and down movement can happen as many as 20 or 25 times.

If you were to break open a hailstone, you would see that it has layers of "skin" from each time it rose and fell in the clouds.

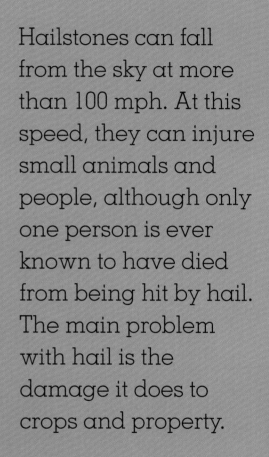

## FAST FACT

Hailstones can fall from the sky at more than 100 mph. At this speed, they can injure small animals and people, although only one person is ever known to have died from being hit by hail. The main problem with hail is the damage it does to crops and property.

# More about hail

Each time the hailstone rises and falls within the
clouds, it gets bigger and bigger. Finally, it gets
so big, it falls to earth. Most hailstones are about
the size of small marbles, but sometimes they
can get much larger. The largest single
hailstone on record weighed more than
$1\frac{1}{2}$ pounds and was bigger than a
softball. Once in India, many hailstones
melted together on the way down
and created a hailstone as big as
a baby elephant!

The people who live in the states of Nebraska, Colorado, and Wyoming know a lot about hail. More hail falls in those states than any others.

While hail usually falls in the summer, sleet usually falls in the winter. Sleet forms when raindrops fall through cold air and freeze. Unlike hail, sleet doesn't get bounced up and down inside a thunderstorm cloud. If there is a layer of warmer air closer to the earth, sleet will often melt and become rain again.

# **Why** do I sometimes see a rainbow after it rains?

After it rains, there is still a lot of water in the air. Rainbows are caused by sunlight shining through and bouncing off water droplets. In fact, if there isn't any rain in the area, you won't see any rainbows.

Rainbows are very colorful. Here's why: As sunlight hits the water droplets in the air, the light is split into seven colors—red, orange, yellow, green, blue, indigo, and violet. A good way to remember the colors of the rainbow is to remember the name Roy G. Biv. The letters in that name are the first letters of the rainbow's colors. Red is always at the top of the rainbow; violet is always at the bottom.

The best time to see rainbows is after an afternoon rain. But you'd better hurry—rainbows usually last only a few minutes.

| R | = | red |
| O | = | orange |
| Y | = | yellow |
| G. | = | green |
| B | = | blue |
| I | = | indigo |
| V | = | violet |

Years ago, some people believed that there was a pot of gold at the end of a rainbow. That's not true, but it's fun to think about.

# More about rainbows

Most of the time, we only see part of a rainbow. That's because houses, trees, or even clouds can get in the way. Other times there aren't enough water droplets to make a rainbow in a part of the sky. The higher up you are, however, the more of a rainbow you'll probably see. Sometimes, it's even possible to see a rainbow in the shape of a circle. To see one of these, you'd have to be in an airplane looking down, watching the sun shine where rain is falling. It's even possible to see a double rainbow in the sky, but they are rare.

You can actually make your own rainbow. All you have to do is wait for a bright, sunny day. With your back to the sun, turn on a garden hose and spray a fine mist of water. Just be sure that you're between the mist and the sun, and you should see a rainbow.

# Why doesn't it rain in the desert?

Actually, most deserts do get some rain. They just don't get very much.

Here's one reason: the areas of the earth near the equator usually get lots of rain. As the rain falls, the air up high becomes cooler and it begins to fall to earth. As it gets closer to earth, the air begins to warm, but without any water in it, this air is very dry. When the air drops down to the earth, it blows hot and dry over certain areas. These areas become deserts.

## HOW LOW CAN YOU GO?

You could go to Death Valley in California and Nevada.

It is 282 feet below sea level . . . the lowest point in the Western Hemisphere.

FAST FACT

In an average year, the city of Cairo, Egypt, gets only about one inch of rain.

If you look on a map, you'll see that many deserts are next to areas that have lots of rainfall. As the air moves toward the equator, it picks up more water droplets and starts rising again. But as long as the air is hot and dry, you can be sure that the ground below will be a desert.

# Why doesn't it rain in the desert?

There's another reason why a desert area may not get much rain: mountains may get in the way.

This usually happens when the mountains are near a large body of water such as an ocean. Here's what happens: As the warm, wet air blows in from the sea onto the land, it bumps into mountains near the shore. The wet air can't go through the mountain, so it is blown up the side. As it rises, the air begins to cool. When it cools enough, the water in the clouds falls on the mountain as either rain or snow. But the water only falls on the side of the mountain facing the wind. That's because the shape of the mountain keeps pushing the air higher and higher until it loses almost all of its water.

Hold it! Rain stays on this side.

On the other side of the mountain, something very different happens. When the cool air crosses the mountain peak, it falls down the back side of the mountain. This falling air doesn't have much water left in it, and as it falls down the mountain, it gets warmer and warmer. After a while, all this warm, dry air blowing on the land behind the mountain turns the land into a desert.

If you look at a map of the west coast of the United States, you'll see an example of this. On the side of the mountain facing the ocean, the land is covered with trees and plants. On the other side of the mountain, the land is desert. Most deserts get less than 10 inches of rain per year, but most trees need at least 30 inches of rain per year to survive. This is why you don't see many trees on the desert. Some cacti and other special plants can survive with as little as 2 or 3 inches of rain per year, so that's why you may see them growing in the desert.

Good thing I'm not thirsty!

# Why does it snow in some places and not in others?

In order for it to snow, there have to be enough tiny droplets of water, called water vapor, in the clouds. Sometimes water vapor collects and forms raindrops, but snow does not start out as rain. For snow to form, the air in the clouds must be cold enough for the water vapor to form crystals. When these snow crystals get heavy enough, they fall to the earth.

I'm made of all these cool-shaped snowflakes!

As snowflakes fall, they drop through several different layers of air. Some of these layers are colder or wetter than others. These layers help shape the snowflakes in different ways.

Not all snowflakes are shaped like stars. Some are shaped like needles and some even look like little capped columns!

You're different, but you're cute!

75

# More about snow

Sometimes, even though snow may be falling from the clouds, the air near the ground is warm enough to melt the snow as it falls. When this happens, the snow turns to rain.

In some places, the air doesn't get cold enough or there isn't enough water vapor in the air to form snow crystals in clouds. These places almost never get snow.

In other places, the snow never melts. It just falls and packs together. In these places, glaciers can form. Glaciers are large areas of packed snow and ice that move slowly until they break apart or melt. Some glaciers last for thousands of years.

Some animals take on special coloring and special habits during snowy weather. Sometimes their coats turn white to help them hide better in the snow. Sometimes they sleep all winter. Sleeping all winter is called hibernation.

The Inuit people have lots of different words for different kinds of snow. They have a word for falling snow and a different word for snow that is piled up. They also have words for packed snow and icy snow.

# **Why** does the weather change at different times of the year?

The weather changes during a year are called seasons. We have seasons because the earth doesn't sit up straight. That may sound like a funny answer, but it's true. If you could get way out in space, you'd see that the top half of the earth is tilted a little more toward the sun during part of its trip around the sun. The sun is a big, fiery ball and you know what happens when you get close to fire . . . Things get hot! This hot weather comes at a time we call summer.

Too hot!

Too cold!

When the top half of the earth is tilted away from the sun a little, things cool off a lot. This is the time we call winter.

That's more like it!

During the spring, the earth is between winter and summer and the weather isn't usually too hot or too cold.

During autumn, or fall, the earth is between summer and winter, and the weather is, again, not too hot or too cold. If you lived on the bottom half of the earth, your weather would be just the opposite. Winters would be nice and warm and your summers would be freezing cold. It takes the earth 365 days to go around the sun and return to its starting point. We call this time period a year.

# More about changing weather

There are three places on the earth where the seasons don't change very much: the top, the bottom, and the middle. The top and the bottom of the earth (called the poles) don't get many direct rays from the sun, so they stay very cold. The middle of the earth (called the equator) always gets direct rays from the sun, so it usually stays very warm there.

It's always cold here.

It's always warm here.

Brrr

Seasons are important because they allow plants time to grow and then rest. And each time the earth makes its trip around the sun (one year), the cycle repeats itself. Spring comes and new plants and animals are born. During the summer they grow and get stronger. In the fall, many plants shed their leaves and the animals begin storing food. In the winter, most plants and animals rest and begin preparing for the new spring time.

Many animals move to warmer areas when the seasons change. This movement is called migration.

# El Niño

An El Niño is an event that happens every 3–7 years. It is the name given to the warming of water near South America. Normally, the water is cool on the east side of the Pacific Ocean by the country of Ecuador. Winds push this cool water toward the western Pacific, and the water begins to warm. Because of these winds, the surface of the water is about 1/2 meter higher in Indonesia than on the coast of South America.

During an El Niño, the air pressure over Indonesia builds up and pushes the warm water eastward toward South America. Because this warming often occurs around Christmas time, this event is called El Niño, which means, "the little boy," after the Christ child.

This change in the water temperature off the coast of South America causes major changes in weather patterns around the world. During an El Niño event, it is common for the southern half of the United States to get much more rain than usual. The western areas of the United States often get less rain than usual. Also, winters in the northern part of the United States are generally milder than normal.

# La Niña

Sometimes the water off the coast of South America becomes unusually cold. When this happens, it is called a La Niña. During a La Niña, ocean temperatures off the coast of South America can fall as much as 4°C (7°F). This may not seem like much, but it makes a big difference in the weather.

HMMM... SEEMS A LITTLE COLDER.

SOUTH AMERICA

 During a La Niña, parts of Australia and Indonesia may receive much more rain than usual. In the United States, a La Niña may make winters warmer in the south and colder in the northwest.

The cooler water begins to cool the air over it. As the air pushes west, it cools all the water over which it passes. Scientists know that cold water deep in the ocean comes to the surface and helps to cause a La Niña, but they are still studying why it happens.

# *Extreme* Weather

**Q.** What is the foggiest place in the United States?

**A.** That would be Cape Disappointment, Washington. If you lived there, you'd be in a fog more than 100 days each year.

**Q.** Where is the wettest, soggiest place on earth?

**A.** The Alakai Swamp in Hawaii. Each year, more than 40 feet of rain falls on this wet spot.

**Q.** Where is the driest place on earth?

**A.** The driest place on earth is in the country of Chile, and it's called the Atacama desert. Don't bother trying to measure annual rainfall here. Some parts of this desert haven't seen any rainfall for hundreds of years.

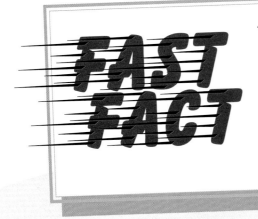
The heaviest snowstorm ever recorded began On February 13, 1959. This snowstorm began at the Mt. Shasta Ski Bowl in California. It lasted seven days. When the storm was over, it had dropped almost 16 feet of snow on the mountain. This is a world record for a single snowstorm.

The most snow in a year fell at the Paradise Ranger Station on Mt. Ranier, Washington, during the 1971–72 snow season, when more than 90 feet of snow fell.

90'

SNOW

WHOA!

# Weird Weather

## Aeolian sounds

When the wind zips through twigs or electrical wires or even as it blows by our ears, it sometimes makes a whistling or humming sound. These sounds are called Aeolian sounds. They are named after the Greek god of winds, Aeolus.

## Ball lightning

Most of the lightning we see looks like an electrical bolt. Sometimes the bolt splits and the lightning "forks." Every so often, lightning appears as a ball. It's usually about a foot in diameter and is often reddish. It may move quickly over objects and has even been known to float in mid-air.

## TRY IT!

There are probably some living thermometers in your backyard! They're called crickets. Count the number of times a cricket chirps in 15 seconds, and then add 37 to that number . . . that's the temperature in degrees Fahrenheit!

# Did **bad weather** kill the dinosaurs?

Scientists are not sure why dinosaurs became extinct, but there's a good chance that weather had something to do with it.

Many scientists think that a large asteroid hit the earth during the time of the dinosaurs. The heat from the asteroid would have created forest fires. The impact would have thrown huge amounts of dust and dirt into the air. That smoke and dust would have blocked much of the sunlight in the atmosphere.

**"I've got a bad feeling about that thing."**

In a very short time, the whole food chain would have been affected. With the sun blocked out, the weather would have become colder and colder. Large dinosaurs with no fur or feathers would have frozen to death. Without enough sunlight, many plants would have died. Plant-eating dinosaurs would have had nothing to eat. And the dinosaurs that ate other dinosaurs would have starved when the plant-eaters died.

# Floods

Which of these weather events do you think kills the most people each year?

- **Hurricanes**
- **Tornados**
- **Lightning**
- **Wind storms**
- **Floods**

The correct answer is **FLOODS!** About 100 people die each year in floods. And floods destroy homes, cars, and other property. Many people don't realize that even shallow water can be very dangerous if it's moving fast. It only takes about two feet of fast-moving water to lift a car, and as little as six inches of fast-moving water can knock a person down.

Floods are usually caused by heavy rain, but they can also be caused by snow melting in mountains or on icy rivers. If the soil is very wet or frozen, sometimes even a normal rainfall can cause flooding. That's because the ground is just too wet or hard to soak up all the water.

**FAST FACT** Fires are the most common natural disaster. Floods are the second most common. In 1889 more than 2,000 people died in Johnstown, Pennsylvania, when a dam broke and flooded the town.

# Here is a list of the ten worst floods ever recorded:

| Date | Location | Deaths |
|---|---|---|
| 1887, September–October | Hwang Ho (Yellow) River, China | More than 900,000 |
| 1939 | North China | 500,000 |
| 1642 | Kaifeng, Honan Province, China | More than 300,000 |
| 1099 | England and the Netherlands | 100,000 |
| 1287, December 14 | The Netherlands | 50,000 |
| 1824 | Russia | 10,000 |
| 1421, November 18 | The Netherlands | 10,000 |
| 1964, November–December | Mekong Delta, South Vietnam | 5,000 |
| 1951, August 6–7 | Manchuria | 4,800 |
| 1948, June | Foochow, China | 3,500 |

# *Ouch!* Sometimes Weather Hurts!

## Chapped lips

Just about everybody has had chapped lips. Our lips get chapped when the skin covering them loses its moisture. Air that is cold or dry can take some of the moisture out of the skin on the lips. Lips can also get chapped when the inside air is very dry. When we're outside during the winter, most of our skin is covered with clothes, but our lips are usually exposed. That's why they get chapped so easily.

## Chilblains

Sometimes when we're out in the cold, our bodies move our warm blood away from the skin and toward our heart and other organs. Our bodies do this to protect us. When this happens, our exposed skin might get red and itchy and it may even crack and swell a little. This is called chilblains. The word comes from two old words that mean to "chill" and "swell." You can see why they picked that name.

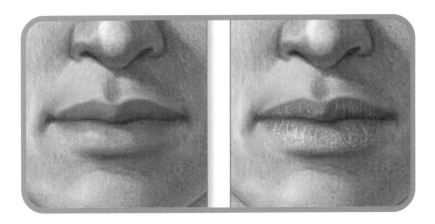

**Lips with plenty of moisture (left) and lips that are chapped (right).**

## Frostbite

Frostbite occurs when skin tissue and blood vessels are damaged from exposure to temperatures below 32 degrees Fahrenheit. It most commonly affects the toes, fingers, earlobes, chin, cheeks, and nose, body parts that are often left uncovered in cold temperatures. Frostbite can occur gradually or rapidly. The speed with which the process progresses depends upon how cold or windy the temperature conditions are and the duration of exposure to those conditions.

## There are three stages of frostbite:

**Frostnip:** You've probably had frostnip if you've been out in the cold and your skin feels like pins and needles are sticking you. Sometimes your skin turns very white and it gets soft. Frostnip is not serious.

**Superficial Frostbite:** This kind of frostbite is more dangerous. Instead of your skin just feeling prickly, it starts to get numb and you can't feel it anymore. Sometimes you might even get blisters. This is dangerous.

**Deep Frostbite:** This is very serious. When a person gets deep frostbite, it isn't just their skin that freezes. Everything under the skin starts to freeze too. If the skin isn't protected immediately, there can be permanent damage.

# Clichés

Sometimes when people talk or write, they use common expressions called clichés. Lots of clichés have to do with weather. Here are some clichés you've probably heard, along with what they mean:

- It's raining cats and dogs ................. *It's raining very hard*

- You're a fair weather friend ............. *You're only a friend in good times.*

- He's skating on thin ice ..................... *He's doing something very risky.*

- It came from out of the blue ............. *Something happened unexpectedly.*

- It's the calm before the storm ......... *Things are quiet now, but that will soon change.*

- She's under the weather ................... *She's not feeling well.*

- You're all wet ........................................ *You're wrong.*

- The sky's the limit ............................... *Anything is possible.*

- Every cloud has a silver lining .......... *There's something good in everything.*

- It happens once in a blue moon ........ *It happens very infrequently.*

# Glossary

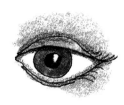

| | |
|---|---|
| **barometer** | An instrument used to measure the weight of air |
| **condensation** | Tiny drops of water that form when warm air meets cold air |
| **funnel cloud** | A dark, fast-spinning cloud that can be seen during a tornado |
| **glaciers** | Large areas of packed snow and ice that move slowly until they melt |
| **jet stream** | Powerful wind more than 6 miles up above the earth |
| **knots** | The measurement of speed throughout the air or water |
| **meteorologist** | A person who is trained to understand and predict the weather |
| **migration** | The seasonal movement of animals to and from warm areas |
| **solar eclipse** | The blocking of the sun's rays caused by the moon moving between the earth and sun |
| **storm surge** | A wall of water built up in front of an oncoming hurricane |
| **ultraviolet rays** | Invisible rays from the sun that can harm skin |
| **visibility** | How well people can see the ground around them |
| **water spout** | A funnel cloud that forms over water |
| **water vapor** | Tiny droplets of water in the air |
| **wind chill** | The cooling effect of wind on skin |
| **wind shear** | A sudden change in the direction of the wind, especially up or down |

# Fun Ways to Use the Glossary

Here are just a few of the many fun games you can play with weather words!

**1.** You can play a Jeopardy-style game: A parent or another child can read the definition. The first person to respond with a question in the format "What is . . . ?" using the correct word wins the round!

**2.** A group of children or children and parents can play "Dictionary": One person chooses a weather word from the glossary and has the correct definition for the weather word. All others make up a definition. Everyone then votes for the definition he or she thinks is the real one. Each correct guess gets 1 point. When all of the glossary terms have been used, the person with the most points wins!

**3.** For younger children, a Pictionary-style game can be a lot of fun: One child chooses a word to draw. Using a timer, he or she tries to draw the weather word so that another child or parent can guess it within the allotted time. Most words guessed makes a winner!

**4.** Or adapt Hangman, 20 Questions, or any other word game that you like!

# Index

# About the Author

**Mike Artell** has written and illustrated many children's books. This book was especially fun for Mike to write and illustrate because he sees so much variety in the weather where he lives. Mike lives in Covington, Louisiana, which is usually warm and humid. However, every ten years or so it snows in Covington. Every few years, hurricanes come close by. Sometimes there are even water spouts on nearby lake Ponchartrain.

In addition to writing and illustrating books, Mike visits schools and libraries around the country and shares his ideas for helping children and teachers think, write, and draw more creatively. You can visit Mike on his website at www.mikeartell.com.